Amer Taqa

The Effect of Newly Prepared Cleansing Agent on The Surface Roughness and Tensile Strength of Highly Impact Acrylic Denture Base Material

Document Nr. V210058

Amer Taqa

The Effect of Newly Prepared Cleansing Agent on The Surface Roughness and Tensile Strength of Highly Impact Acrylic Denture Base Material

GRIN Verlag

1. Auflage 2012
Copyright © 2012 GRIN Verlag GmbH
http://www.grin.com
Druck und Bindung: Books on Demand GmbH, Norderstedt Germany
ISBN 978-3-656-39054-1

The Effect of Newly Prepared Cleansing Agent on The Surface Roughness and Tensile Strength of Highly Impact Acrylic Denture Base Material

Amer A Taqa , Ammar Kh. Al-Noori and Hasan k.Al-Omari, Department of Prosthetic Dentistry, College of Dentistry, University of Mosul

Abstract

Aims: To evaluate the effect of two prepared and commercial solution on surface roughness and tensile strength of highly impact acrylic denture base material **Materials and Methods:** The total number of specimens were one hundred and fifty. They were prepared from highly impact acrylic and subdivided into five groups for each solution (EDTA, Soda+H_2O_2, Lacalut, Corega and distilled water). Two laboratory tests were used for this research. Surface Roughness and tensile strength test. The surface roughness test specimens were constructed with dimensions ($10\times10\times2\pm0.03$mm) (length, width, and thickness respectively). According to ADA specification no.12 the tensile test specimens were constructed with dimensions $90\times10\times3\pm0.03$mm (length, width, and thickness respectively). The immersion periods in this research are (2day, 7 day and one month). ANOVA and Duncan multiple range test were used. The statistical results were considered significant at p $\leq$ 0.05. **Results:** the results showed that (soda+H_2O_2) has no significant change on the surface roughness and tensile strength of highly impact acrylic denture base material in (2day, 7 day, and 1month) **conclusion:** (soda+H_2O_2) has the lowest effect on the surface roughness and tensile strength of highly impact acrylic denture base material in (2day, 7 day, and 1month)

Key words: denture cleanser, highly impact, lacalut

Introduction

Polymethyl methacrylate (PMMA) has been used in dental prosthetic devices for almost 70 years. Three fundamental features have contributed for its success: excellent appearance, simple processing technique and easiness of the repair. However, the resistance to impact and fracture of PMMA during function are low fracture [1-3]. The denture base resin is subjected to various stresses during function. During fabrication of a denture, the physical and mechanical properties influence by cure condition and choice of materials. Each cure cycle or fabrication technique is a compromise that attempts to optimize the properties thought important for a given application. Dentist and manufacturers of denture base materials have long been searching for ideal materials and designs for dentures. So far, the results have been noteworthy, although there are still some physical and mechanical problems with these materials [4].Many attempts have been made to enhance the strength properties of acrylic denture bases including the addition of metal wire. The primary problem of using metal wire reinforcement is poor adhesion between wire and acrylic resin. Although several methods have been used to improve the adhesion between these components, enhancement in mechanical properties, such as transverse strength and fatigue resistance, was not significant. [5,6] Modifications of chemical structure, by the addition of cross-linking agents such as polyethyleneglycol di-methacrylate or by copolymerization with rubber, have been attempted [7] Various types of fiber including carbon fiber whisker fiber, aramid fiber, polyethylene fiber, and glass fiber have been used as a reinforcement. Reinforcement with fibers enhances the mechanical strength characteristics of denture bases, such as the transverse strength, ultimate tensile strength and impact strength. In addition, fiber reinforcement has advantages compared with other reinforcement methods, including improved esthetics, enhanced bonding to the resin matrix, and ease of repair [8-12] Cleansers and cleaning methods used may have harmful effect on the plastic or metal component of the denture. Knowledge of constituents of denture cleansers, their efficiency, adverse effect and safety would aid in dispensing appropriate information to the patient , so the dentist must be able to recommend a denture cleanser that is effective, non deleterious to denture material and safe for patient use. [13,14]

During this resarch of the effect of denture cleanser on the properties of denture base materials, the chemicals disinfectants (Chlorhexidine gluconate, sodium hypochlorite and gluteraldehyde) reduced the tensile strength of denture base material, but this reduction is not significant. [15] The aims of this study are to evaluate the effect of two prepared and two commercial solutions on surface roughness and tensile strength of highly impact acrylic denture base material after (2day, 7 day and 1 month).

Materials and Methods

The total number of specimens was one hundred and fifty. Seventy five for each test were prepared from highly impact acrylic and subdivided into five groups for each solution. The immersion periods in this study are (2day, 7 day and one month)

Highly impact acrylic (vertex-dental) used in this research mixed according to the manufacture instruction. The liquid powder ratio is 1 ml liquid and 1.2 mg powder, adding powder to the liquid and then mixing the powder to liquid for 30 min , leave the mixing for 8 min in room temperature 22 °C until reach to the dough stage adding the highly impact acrylic to the flask through in room temperature 22 °C and then press the flask by press , and putting immediately inside hot water approximately 70°C for 90 min and then rising the degree of temperature to the 100 °C for 30 min and the remove the flask and leave it to cool. Two laboratory tests were used for this research. tensile strength and Surface roughness test, According to ADA specification no.12 the tensile test specimens were constructed with dimensions $90\times10\times3\pm0.03$mm (length, width, and thickness respectively) Then universal testing machine (Gunt, Germany) was used to measure the tensile strength of specimens The force at failure was recorded in Newton (N) and the true tensile strength value was calculated by the following formula: Tensile strength = F(N)/A (mm^2) [16] . The surface roughness test specimens were constructed with dimensions ($10\times10\times2\pm0.03$mm) (length, width, and thickness respectively) The surface roughness (Ra) values were measured using a profilometer (Stylus 10 UK) which can measure small surface variations by moving a diamond stylus in contact with the surface.[17]. The specimens were fabricated by using Type III model dental stone (Zhermack SPA Rovigo, Italy) as a mold. This study deals with five solutions (table 1).two experimental prepared solutions, solution one

(Ethylene Diamin Tetra acetic Acid) EDTA and solution two (soda Na_2Co_3 and Hydrogen peroxide H_2O_2) two commercial denture cleanser tablets (Corega, lacalut) for comparison and distilled water as a control solution. Every solution was diluted in 100 ml of distilled water.

The following equations illustrate the preparation of the above solutions

1-EDTA

2-Soda +H2O2

$$H_2O_2 Na_2CO_3 \longrightarrow Na_2HCO_3 + CO_2 + 1/2O_2$$

Artificial Saliva was developed in order to bring the trials closer to real in-mouth conditions. Indeed, its mineral composition is close to that of resting mixed saliva.

By mixing the following compounds in distilled water, the artificial saliva solution were prepared [18].

Compounds	Concentrations (mg/L)
➢ NaCl	0.4
➢ KCl	0.4
➢ $CaCL_2$	0.79
➢ NaH_2PO_4	0.78
➢ UREA	1
➢ DISTELD WATER	1 L

The fresh solutions were prepared daily at the beginning of soaking trial (1/2h). The specimens were removed from the solution washed with distilled water, and dried in air by shaking the specimen for about 30 seconds. The solutions were removed, the beakers were cleaned and the specimens were immersed in distilled water for 8 hs at (21±2°C) then immersed in artificial saliva for about 15.5 h at (37±1°C)in the incubator. According to method described previously The immersion periods in this study are (2day, 7 day and one month). [19].

Lacalut denture cleanser, release an active oxygen, and Corega denture cleanser, release an active CO2, used in this study and prepared as manufacture instruction

The following statistical methods were used to analyse and assess the results via SPSS V. 11.5 for Windows:

1. Descriptive statistics include mean ± standard deviation values.

2. ANOVA and Duncan multiple range test were used. The statistical results were considered significant at $p \leq 0.05$.

Results

Tensile Strength

The One Way Analysis of variance (ANOVA) as shown in Tables (2) demonstrated that there was significant difference at $P \leq 0.05$ in the tensile strength of highly impact acrylic resin in 2day and 7 day and no significant difference in 1 month. Figures (1-3) demonstrated the mean ± SD values and Duncan's multiple range test of tensile strength. In 2 day showed the highest value in (EDTA) and lowest value in (Distilled water) .in 7 day showed the highest value in (Corega) and the lowest value in (Lacalut)

The One Way Analysis of variance (ANOVA) as shown in Table (3) demonstrated the differences between the solution and showed that there was significant difference at $P \leq 0.05$ in the tensile strength of highly impact acrylic resin in (EDTA, Distilled water,Corega, Lacalut) and no significant difference in solution $(Soda+H_2O_2)$.Tensile strength of highly impact acrylic resin, in comparison between solution, figures (4-8) demonstrated the mean ± SD values and Duncan's multiple range test of tensile strength. In EDTA) showed the highest value in 2 day and lowest value in 1 month .In $(Soda+H_2O_2)$ showed no differences. In (Distilled water) the highest value in 7 day and lowest value in 1 month .In (Corega) showed the highest value in 7 day and lowest value in 1month in (Lacalut) showed the highest value in 2 day and lowest value in1month

Surface Roughness

The One Way Analysis of variance (ANOVA) as shown in Table (4) demonstrated that there was significant difference at P≤ 0.05 in the surface roughness of highly impact acrylic resin among time intervals. Surface roughness of highly impact acrylic resin, in comparison between time intervals, figures (9-11) demonstrated the mean ± SD values and Duncan's multiple range test of surface roughness. . In 2 day showed the highest value in (Distilled water) and lowest value in (Lacalut) .In 7 day showed the highest value in (Lacalu)t and the lowest value in (Corega). In 1 month showed highest value in (Lacalut) and the lowest value in (Soad+H_2O_2).

The One Way Analysis of variance (ANOVA) as shown in Table (5) demonstrated that there was a significant difference at P≤ 0.05 in the surface roughness of high impact acrylic in solutions (EDTA, distilled water,corega,lacalut) and no significant difference in (Soda+H_2O_2). Surface roughness of highly impact acrylic resin, in comparison between five solutions, figures (12-16) demonstrated the mean ± SD values and Duncan's multiple range test of surface roughness. In (EDTA) showed the highest value in 7 day and lowest value in 2day. In (Soda+H_2O_2) showed no differences. In (distilled water) the highest value in 1 month and lowest value in 2day. In (Corega) showed the highest value in 2 day and lowest value in 1month in (Lacalut) showed the highest value in 1 month and lowest value in 2day

Discussion

(Ethylene Diamen Tetra acetic Acid) EDTA show significant change in surface roughness and tensile strength in 2 days,7 days and one month because it release CO_2 and form weak sodium bicarbonate that effect on the surface roughness of highly impact acrylic denture base material[20] . **(Soda+H_2O_2)** show there was no significant change in surface roughness and tensile strength in 2 days,7 days and one month because it release O_2 which the different reaction from other solution and it was not effect on the surface roughness of highly impact acrylic denture base material[21] .**(Distilled water)** show significant change in surface roughness and tensile strength in 2 days,7 days and one month that agreement with Pavarina et al[22] stated that prolonged immersion of denture teeth in water caused softening of the acrylic resin. Absorbed water has been shown to affect the surface properties of all

forms of acrylic [23] (**Lacalut denture cleanser**) show significant change in surface roughness and tensile strength in 2 days, 7 days and one month because it release CO_2 and O_2 and form weak sodium bicarbonate that effect on the surface roughness of highly impact acrylic denture base material   this result disagreement with other studies Salman and Saleem [24] show that there was no significant difference between pre and post soaking for heat cured acrylic group. (**Corega**) show significant change in surface roughness and tensile strength in 2 days, 7 days and one month because it release CO_2 and form weak sodium bicarbonate that affect on the surface roughness of highly impact acrylic denture base material that disagreement with Ural.[25] show the test specimens that immersed in water with Corega Tabs did not show a significant increase in surface roughness of the specimens. FDA is asking manufacturers of denture cleansers to include a warning in the label about persulfates, which are known to cause allergic reactions in some people. Persulfates are used in most denture cleansers as part of the cleaning and bleaching process. The agency is also recommending that manufacturers consider appropriate alternatives to persulfates. the use (EDTA and soda+H_2O_2) not cause allergic reactions because this solution not have Persulfates and it were more safe than other solution[26]

Conclusion

Soda+H_2O_2 had no significant difference in surface roughness and tensile strength in (2day, 7day, and 1month)

In regarding the type of solutions there was no significant difference in tensile strength after (1 month) immersion period

There was a significant difference among the solution surface roughness after difference immersion period (2day, 7 day and 1month)

Table (1) Solutions Preparation

Solution no.	Material 1	Weight or volume	Material 2	Weight or volume
1	EDTA	4 g		
2	Soda		H2O2	25%
3	Distilled Water	100 ml		
4	Corega	1 tab = 3.25 g		
5	Lacalut	1 tab= 2.85		

Table (2)ANOVA for Comparison of tensile strength among time intervals

Time		SS	df	MS	F–value	*p*–value
	Between Groups	0.001	7	0.000	8.013	0.000*
2 day	**Within Groups**	0.001	32	0.000		
	Total	0.002	39			
	Between Groups	0.001	7	0.000	3.434	0.007*
7 day	**Within Groups**	0.002	32	0.000		
	Total	0.003	39			
	Between Groups	0.001	7	0.000	1.187	0.338
1 month	**Within Groups**	0.003	32	0.000		
	Total	0.004	39			

SOV: Source of variance; SS: Sum of Squares; df: Degree of freedom; MS: Mean Square.* indicated significant difference at $p \leq 0.05$.

Table (3)ANOVA for Comparison of tensile strength among five solutions.

Solution		SS	df	MS	F–value	*p*–value
	Between Groups	0.001	2	0.001	12.793	0.001*
EDTA	**Within Groups**	0.001	12	0.000		
	Total	0.002	14			
Soda	**Between Groups**	0.000	2	0.000	2.358	0.137
+	**Within Groups**	0.001	12	0.000		
H_2O_2	**Total**	0.002	14			
	Between Groups	0.001	2	0.000	14.713	0.001*
Corega	**Within Groups**	0.000	12	0.000		
	Total	0.001	14			
	Between Groups	0.001	2	0.000	8.397	0.005*
Lacalute	**Within Groups**	0.000	12	0.000		
	Total	0.001	14			
	Between Groups	0.001	2	0.000	8.895	0.004*
Water	**Within Groups**	0.000	12	0.000		
	Total	0.001	14			

SOV: Source of variance; SS: Sum of Squares; df: Degree of freedom; MS: Mean Square

* indicated significant difference at $p \leq 0.05$.

Table (4)ANOVA for Comparison of surface roughness among time intervals.

Time		SS	df	MS	F–value	p–value
2 day	**Between Groups**	5.057	7	0.722	7.444	0.000*
	Within Groups	3.105	32	0.097		
	Total	8.162	39			
7 day	**Between Groups**	7.098	7	1.014	21.590	0.000*
	Within Groups	1.503	32	0.047		
	Total	8.600	39			
1 month	**Between Groups**	14.257	7	2.037	27.542	0.000*
	Within Groups	2.366	32	0.074		
	Total	16.623	39			

SOV: Source of variance; SS: Sum of Squares; df: Degree of freedom; MS: Mean Square

* indicated significant difference at $p \leq 0.05$

Table (5)ANOVA for Comparison of surface roughness among five solutions.

Solution		SS	df	MS	F–value	p–value
EDTA	**Between Groups**	3.467	2	1.734	20.260	0.000*
	Within Groups	1.027	12	0.086		
	Total	4.494	14			
Soda + H_2O_2	**Between Groups**	1.445	2	0.722	3.186	0.078
	Within Groups	2.720	12	0.227		
	Total	4.165	14			
Corega	**Between Groups**	0.571	2	0.285	14.189	0.001*
	Within Groups	0.241	12	0.020		
	Total	0.812	14			
Lacalute	**Between Groups**	8.352	2	4.176	66.639	0.000*
	Within Groups	0.752	12	0.063		
	Total	9.104	14			
Water	**Between Groups**	0.392	2	0.196	5.751	0.018*
	Within Groups	0.409	12	0.034		
	Total	0.801	14			

SOV: Source of variance; SS: Sum of Squares; df: Degree of freedom; MS: Mean Square

* indicated significant difference at $p \leq 0.05$

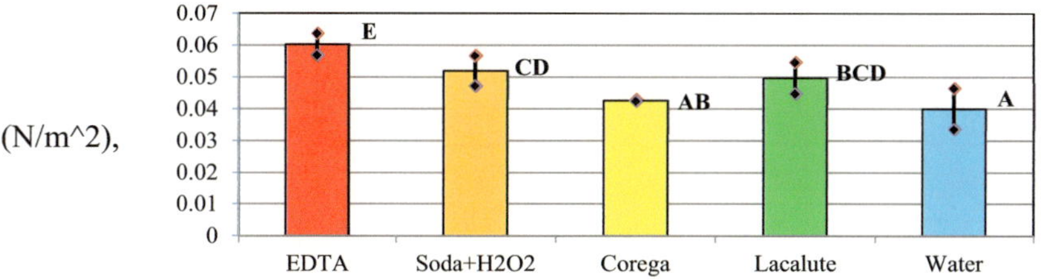

Means with the different letter are statistically significant different at $p \leq 0.05$

Figure (1) Mean ± SD and Duncan's multiple rang test of tensile strength for Comparison among time intervals. (2 Days)

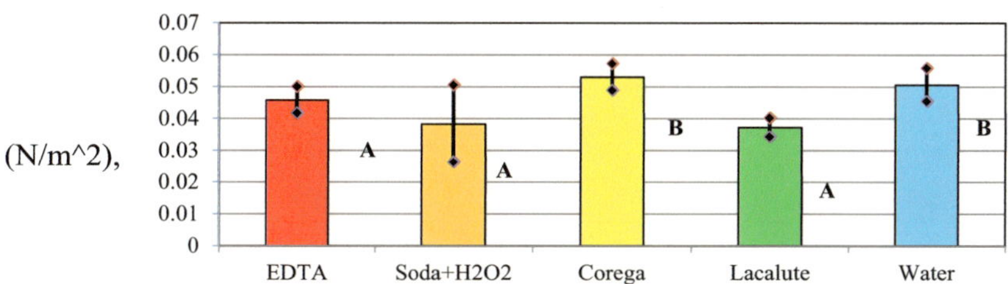

Means with the different letter are statistically significant different at $p \leq 0.05$

Figure (2) Mean ± SD and Duncan's multiple rang test of tensile strength for Comparison among time intervals. (7 Days)

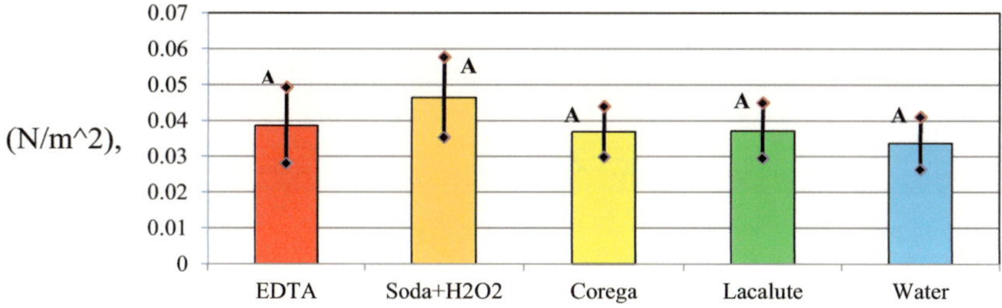

Means with the same letter are statistically no significant different at $p \leq 0.05$

Figure (3) Mean ± SD and Duncan's multiple rang test of tensile strength for Comparison among time intervals.(1 Month)

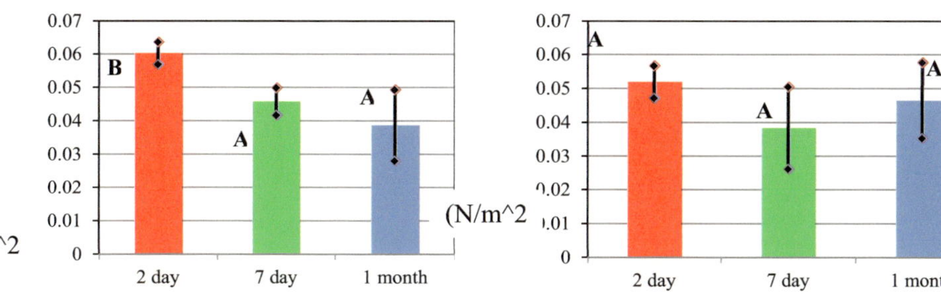

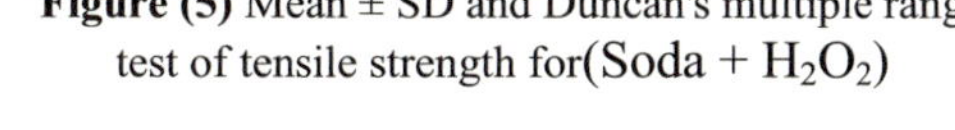

Means with the different letter are statistically significant different at $p \leq 0.05$

Figure (4) Mean ± SD and Duncan's multiple rang test of tensile strength for EDTA

Means with the same letter are statistically no significant different at $p \leq 0.05$

Figure (5) Mean ± SD and Duncan's multiple rang test of tensile strength for $(Soda + H_2O_2)$

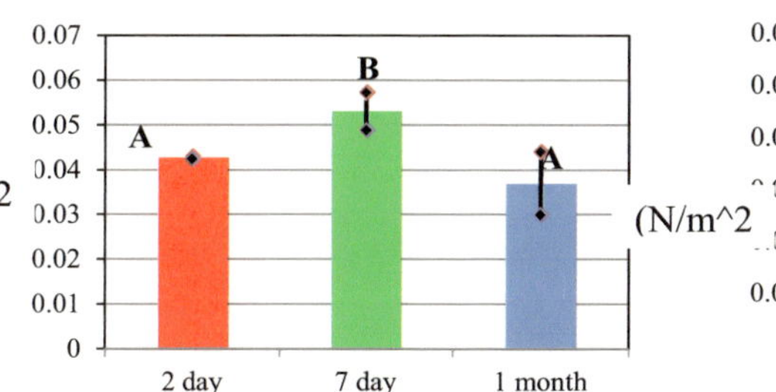

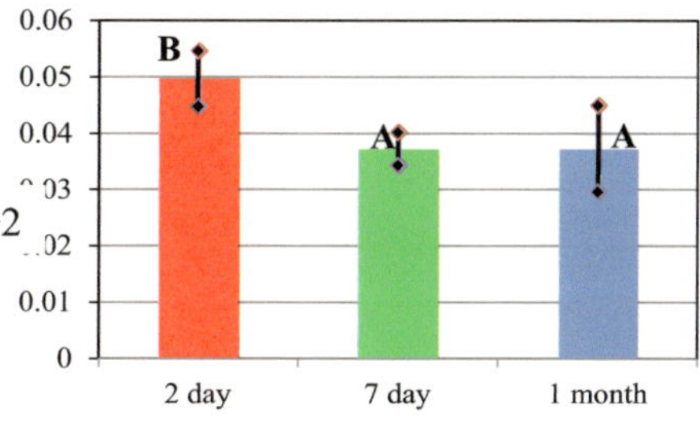

Means with the different letter are statistically significant different at $p \leq 0.05$

Figure (6) Mean ± SD and Duncan's multiple rang test of tensile strength for (Corega)

Means with the different letter are statistically significant different at $p \leq 0.05$

Figure (7) Mean ± SD and Duncan's multiple rang test of tensile strength for (Lacalute)

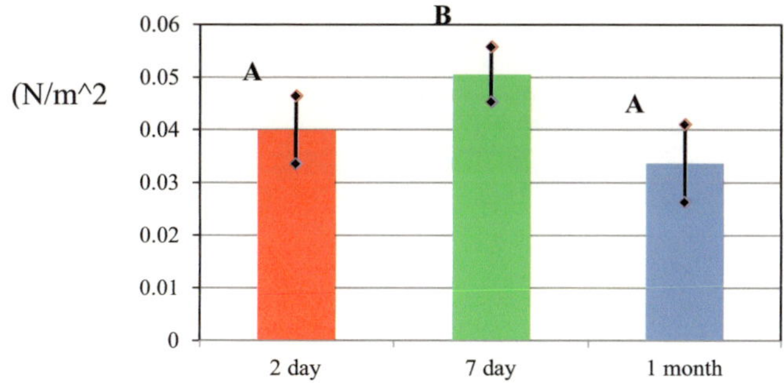

Means with the different letter are statistically significant different at $p \leq 0.05$

Figure (8) Mean ± SD and Duncan's multiple rang test of tensile strength for (Water)

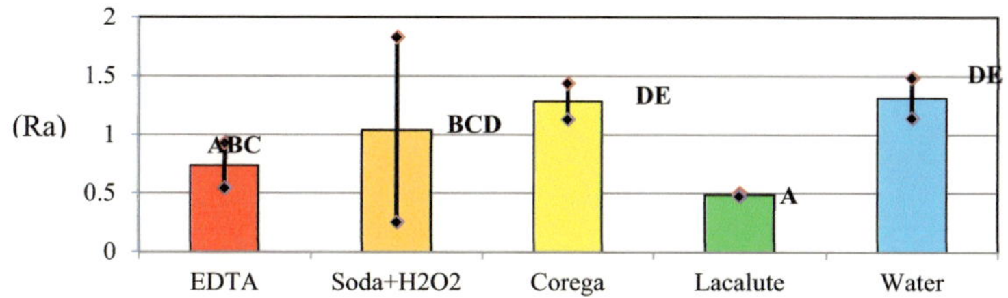

Means with the different letter are statistically significant different at $p \leq 0.05$

Figure (9) Mean ± SD and Duncan's multiple rang test of surface roughness for Comparison among time intervals. (2 Days)

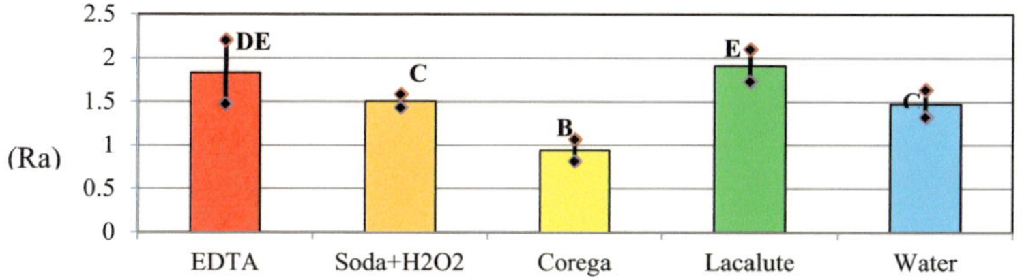

Means with the different letter are statistically significant different at $p \leq 0.05$

Figure (10) Mean ± SD and Duncan's multiple rang test of surface roughness for Comparison among time intervals. (7 Days)

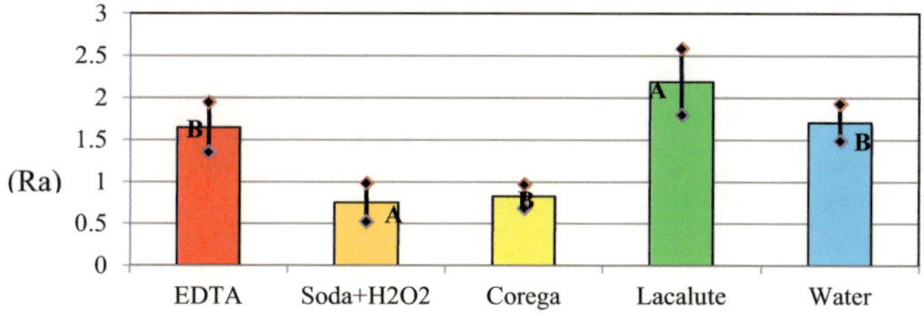

Means with the different letter are statistically significant different at $p \leq 0.05$

Figure (11) Mean ± SD and Duncan's multiple rang test of surface roughness for Comparison among time intervals. (1 Month)

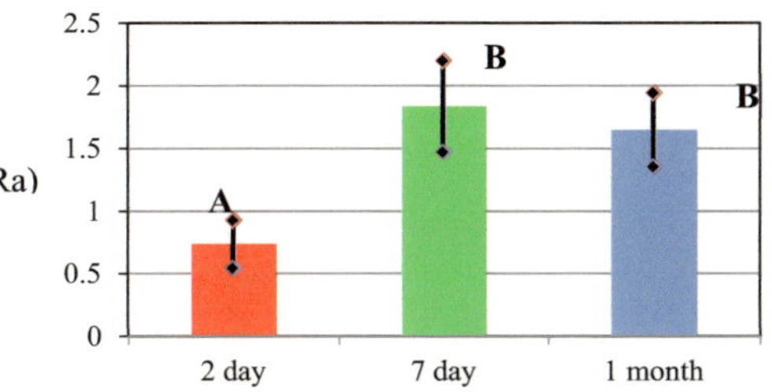

Means with the different letter are statistically significant different at $p \leq 0.05$

Figure (13) Mean ± SD and Duncan's multiple rang test of surface roughness for

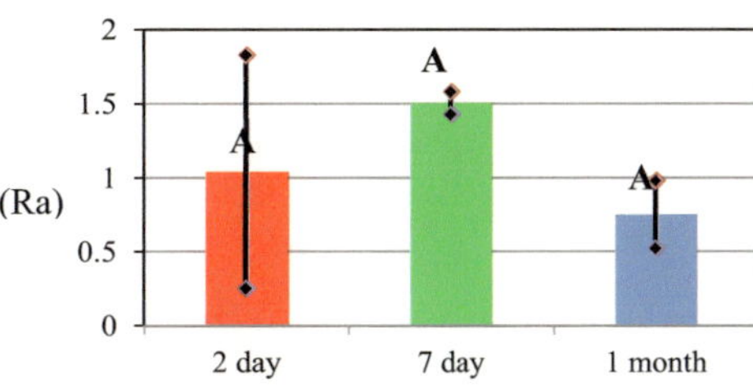

with the same letter are statistically no significant different at $p \leq 0.05$

Figure (12) Means Mean ± SD and Duncan's multiple rang test of surface roughness for (EDTA)

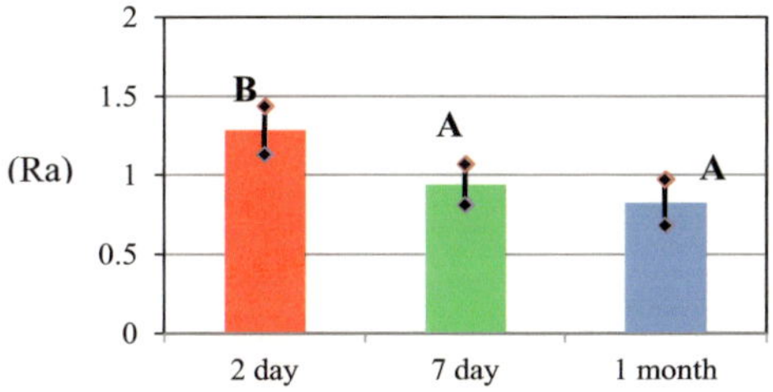

Means with the different letter are statistically significant different at $p \leq 0.05$

Figure (14) Mean ± SD and Duncan's multiple rang test of surface roughness for (Corega)

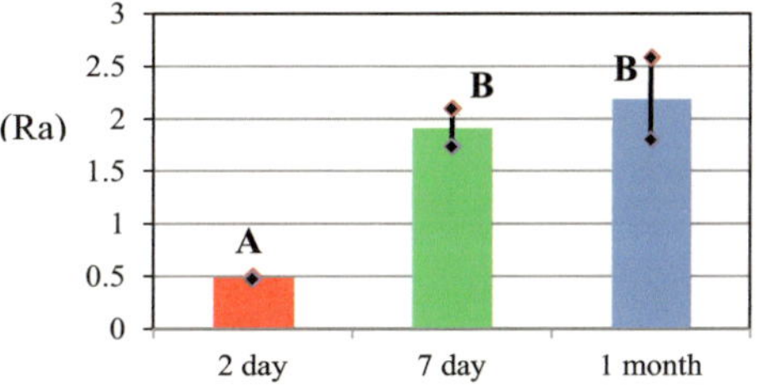

Means with the different letter are statistically significant different at $p \leq 0.05$

Figure (15) Mean ± SD and Duncan's multiple rang test of surface roughness for (Lacalute)

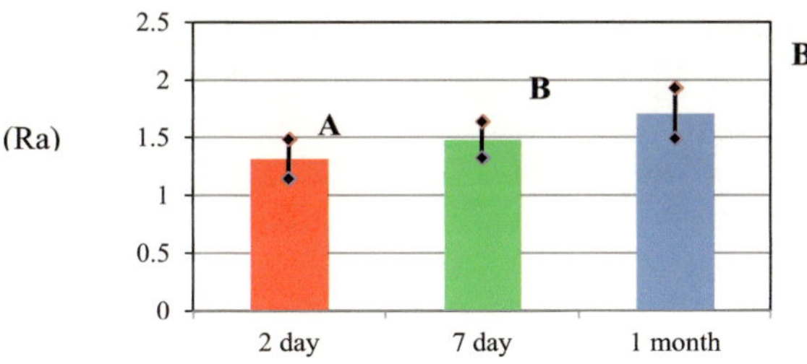

Means with the different letter are statistically significant different at $p \leq 0.05$

Figure (16) Mean ± SD and Duncan's multiple rang test of surface roughness for (Water)

References

1- Pires-de-Souza FC, Panzeria H, Vieira M A , Garcia LF R, Consani S. Impact and fracture resistance of an experimental acrylic polymer with elastomer in different proportions.; *Materials Rese*, 2009: 12(1): 415-418.

2- Bertassoni L, Marshall G, de Souza E and Rached R. Effect of pre- and postpolymerization on flexural strength and elastic modulus of impregnated, fiberreinforced denture base acrylic resins. *J Prosthet Dent* 2008;100:449-457

3- -Pfeiffer p, Rosenbauer E: Residual methyl methacrylate monomer, water sorption, and water solubility of hypoallergenic denture base materials. *J Prosthet Dent* .2004;92:72-8.

4- Mohamed SH, Al-Jadi AM, Ajaal T: Using of HPLC. analysis for evaluation of residual monomer content in denture base material and their effect on mechanical properties. *J Phys Sci*. 2008; 19;2: 127 – 135.

5- kim S and Watts D. The effect of reinforcement with woven E-glass fibers on the impact strength of complete dentures fabricated with high-impactacrylic resin. *J Prosthet Dent*. 2004; 91:274-80.

6- Foo S, Lindquist T, Aquilino S, Schneider R, Williamson D,and Boyer D.Effect of Polyaramid Fiber Reinforcement on the Strength of 3 Denture Base Polymethyl Methacrylate Resins .*J of Prosthodont*. 2001;10:148-153.

7- Robinson, McCabe JF: Impact strength of acrylic resin denture basematerials with surface defects. *Dent Mater*. 1993; 9:355-60.

8- Uzun G, Hersek N, Tincer T. Effect of five woven fiber reinforcements on the impact and transverse strength of a denture base resin. *J Prosthet Dent*.1999; 81:616-20.

9- Karacaer O, Polat T, Tezvergıl A, Lassıla L,and Vallıttu P . The effect of length and concentration of glass fibers on the mechanical properties of an injection- and a compression-molded denture base polymer. *J Prosthet Dent* 2003; 90:385-93.

10-Diaz-Arnold A, Vargas M, Shaull K, Laffoon J,and Qian F . Flexural and fatigue strengths of denture base resin. *J Prosthet Dent* 2008;100:47-51

11-Cheng Y, Phil M,and Chow *T.* Fabrication of Comdete Denture BasesReinforced with Pofyethylene Woven Fabric. *J Prosthod* 1999;8:268-272

12-Vallittu P: Dimensional accuracy and stability of polymethyl methacrylatereinforced with metal wire or with continuous glass fiber. *J prosthod dent.* 1996;75:617-21

13-Kulak-Ozkan Y, Kazazoglu E, Arikan A: Oral hygiene habits, Denture cleanliness, Presence of yeast and stomatitis in elderly people. *J Oral Rehabil.* 2002; 29:300-304.

14-Salem s and Al-Khafaji A : The effect of denture cleansers on surface roughness and microhardness of stained light cured denture base material .*J Bagh Coll Dentistry* 2007; 19;1:1-5

15-Peracini A, Davi L, Ribeiro N, Souza R, Silva C, Paranhos H: Effect of denture cleansers on physical properties of heat-polymerized acrylic resin *Journal of Prosthodontic Research.* 2010;54 ;8:78–83

16--Abdulrahman Omer M. F :The effect of chemical disinfectants on some properties of Valplast and Acrylic denture base materials. Msc thesis . college of dentistry,Hawler Medical University 2011

17-Ozkan YK, Sertgoz A and Gedik H : Effect of thermocycling on tensile bonding strength of six silicone-based, resilient denture liners. *J Prosthet Dent*; 2003:89: 303-310.

18-Nevzatoĝlu E, Ozcan M, Ozkan YK and Kadir T: Adherence of Candida albicans to denture base acrylics and silicone-based resilient liner materials with different surface finished. *J Clin Oral Inves.*2007 ;11;3: 231-237.

19-Neppelenbroek KH, Pavarina AC, Spolidorio DM, :Effectiveness of microwave sterilization on three hard Chair side reline resins. *Int J Prosthodont*.2003; 16:616-620.

20-Giacomelli L, Salerno M, Derchi G, Genovesi A, Paganin PP, Covani U: Effect of air polishing with glycine and bicarbonate powders on a nanocomposite used in dental restorations: an in vitro study. *Int J Periodontics Restorative Dent.* 2011;31:5:51-6

21-Fujiwara .Y , Toyoda N. , Mochiji K, T. Mitamura , I. Yamada Reduction of surface roughness by Ta2O5 film formation with O2 cluster ion assisted deposition. *Nucl. Instr. and Meth. in Phys. Res.* 2003;206; 870–874

22-Pavarina AC, Vergani CE, Machado AL, Giampaolo ET, Teraoka MT. The effect of disinfectant solutions on the hardness of acrylic resin denture teeth. *J Oral Rehabil* 2003; 30: 749-752.

23-Devlin H, Kaushik P. The effect of water absorption on acrylic surface properties. *J Prosthodont.* 2005; 14: 233-238.

24-Salman m and Saleem S; Effect of different denture cleanser solutions on some mechanical and physical properties of nylon and acrylic denture base materials.. *J Bagh Coll Dentistry*.2011; 23:19-24.

25-Ural C ;Effect of different denture cleanser on surface roughness of denture bas materials;clinical dentistry and research.2011 ; 35;2: 14-20

26- *www.fda.gov/consumer*